THE MAUI FIRE DISASTER

Unraveling the Devastating Maui Wildfires

Steve K. Bryant

Copyright

Copyright © 2024 by **Steve K. Bryant**

All rights reserved. No part of this publication may be reproduced, distributed, or transmitted in any form or by any means, including photocopying, recording, or other electronic or mechanical methods, without the prior written permission of the publisher, except in the case of brief quotations embodied in critical reviews and certain other noncommercial uses permitted by copyright law.

Contents

Introduction 6
   An Overview of August 8, 2023's Devastating Maui Wildfires 6
   Paying attention to the Lahaina Fire, which is the deadliest US wildfire in more than a century. 7
   Consequences of Turning Down Further Assistance and the Ensuing Probe 8

Chapter one 11

Background Information 11
   Reports from the Fire Safety Research Institute and Hawaii's Attorney General Anne Lopez 11
   Conclusions drawn from the Western Fire Chiefs Association Study 13
   The Historical Background of Maui's Fire Danger Warnings and Emergency Preparedness 15

Chapter two 18

Timeline of Events 18
   Conditions and Warnings Before a Fire 18
   Initial Wildfire Activity and Evacuation Efforts 20
   Events of the Day, Including Flare-ups and Fire Containment 22
   Obstacles Emergency Responders Face Because of Strong Winds and Inadequate Communication 23

Chapter three 26

Response and Coordination Efforts 26
   Maui Emergency Management Agency (MEMA) and County Officials' Reaction 26
   Requests for Extra Help and Interaction with State Organizations 28
   Resources and Coordination Limited in the Face of Increasing Fire Situations 30

Chapter four 32

Impact and Aftermath 32
   The Lahaina Fire's devastation, which included fatalities and property damage 32

Firefighters' Struggles Fighting the Quickly
Spreading Flames 34

Attorney General Anne Lopez and Mayor
Richard Bissen of Maui County made
statements. 36

Federal agencies' ongoing legal actions and
investigations 38

Chapter five 41

Conclusion 41

An overview of the circumstances leading up
to the wildfires in Maui and their effects 41

Consequences for Upcoming Emergency
Planning and Organizing 43

The significance of comprehending the
disaster's causes and lessons learned 45

Introduction

An Overview of August 8, 2023's Devastating Maui Wildfires

The wildfires that ravaged Maui on August 8, 2023, will go down in the annals of natural disasters as a sobering reminder of both the power of nature and the frailty of humankind. A path of devastation was left in its wake as what started out as a few seemingly harmless brush fires swiftly grew into an unprecedentedly large and destructive conflagration. In addition to claiming lives and livelihoods, the events of that dreadful day revealed structural shortcomings in emergency response and coordination, which prompted a national conversation about the reality of climate change and disaster preparedness.

Paying attention to the Lahaina Fire, which is the deadliest US wildfire in more than a century.

One fire, the Lahaina fire, became the focal point of the tragedy amidst the turmoil and devastation caused by the Maui wildfires. Started by a swiftly spreading brush fire close to Ku-ialua Street and Ho'okahua Place, this fire swiftly spread to hundreds of Lahaina residences and businesses, turning the once-beautiful town into a scene of complete destruction. Fueled by strong winds and dry weather, flames raced through the streets, forcing inhabitants to run for their lives as firemen battled mightily to contain the fire.

The fact that the Lahaina fire is the worst US wildfire in more than a century is evidence of the severity of the incident and its lasting effects on the Maui population. Its legacy serves as a melancholy reminder of the incredible strength of nature and the frailty of human existence, with 101 lives lost and $6 billion in damages. However, there is a deeper story of bravery, resiliency, and the undying spirit of people who survived the storm that lies beyond the startling figures.

Consequences of Turning Down Further Assistance and the Ensuing Probe

Following the Maui wildfires, there are many unanswered concerns regarding the choices taken in the crucial hours preceding the tragedy. The most significant of them is the disclosure that officials disregarded further assistance before to the Lahaina fire's outbreak, even though they were alerted to an increased risk of fire and impending catastrophe. After a thorough inquiry, this decision is currently being scrutinized, which raises serious concerns about the effectiveness of emergency preparedness and the responsibility of those entrusted with ensuring the welfare of the public.

The consequences of turning down more assistance become more and clearer as the investigation progresses. It emphasizes the necessity of increased alertness and preventative actions in the face of growing climate risks, as well as the significance of accountability and openness in disaster response operations. A nation searching for answers struggles with the difficulties of catastrophe management and the necessity of taking lessons from the past to avert tragedies in the future.

We will go deeper into the events of August 8, 2023, revealing the behind-the-scenes tale of the Maui wildfires and examining the effects of this unparalleled calamity on people in the pages that follow. We will see firsthand the tragedies and victories that occurred in the midst of the fires, from the terrifying testimonies of survivors to the unceasing efforts of first responders. Above all, though, we will address the harsh facts of climate change and the pressing need for coordinated action to create a more compassionate and resilient world in the face of misfortune.

Chapter one

Background Information

Reports from the Fire Safety Research Institute and Hawaii's Attorney General Anne Lopez

Following the terrible August 8, 2023, wildfires on Maui, Hawaii Attorney General Anne Lopez launched a comprehensive investigation to determine the circumstances leading up to the disaster. Her agency worked in tandem with the Fire Safety Research Institute to disentangle the intricate series of events that preceded the disaster and illuminate the structural shortcomings that added to its severity. The Fire Safety Research Institute and Attorney General Lopez's studies provide a thorough analysis of the variables at play before to, during, and following the wildfires. These studies offer a thorough picture of the disaster through in-depth examination of official records, eyewitness accounts, and professional insights.

One of the primary findings of the study is that more help was turned down prior to the Lahaina fire starting, despite warnings of an elevated fire hazard. Based on the evaluations, this decision raised significant doubts regarding the efficacy of the preventative actions and had a significant impact on emergency response activities.

The investigations also brought to light the challenges first responders faced, like inadequate communication and limited resources, in trying to contain the rapidly expanding fires. By documenting the sequence of events that led up to the catastrophe, Attorney General Lopez and the Fire Safety Research Institute have significantly illuminated the causes of the Maui wildfires and the lessons that may be learned from them.

Conclusions drawn from the Western Fire Chiefs Association Study

The Western Fire Chiefs Association looked investigated the Maui wildfires independently in addition to Attorney General Anne Lopez's investigations. The association's results, which are tasked with evaluating emergency planning and response actions, provide more insights into the variables influencing the disaster's severity.

The Western Fire Chiefs Association inquiry yielded several important conclusions, including worries about coordination issues and failures in communication between local, state, and federal organizations. The research focuses on situations when important information was delayed in getting to decision-makers, which hampered response efforts and made the crisis worse.

The association's inquiry also emphasizes how emergency responders in wildfire-prone areas require better training and supplies. The research highlights deficiencies in readiness and reaction capacities and offers insightful suggestions for enhancing resilience and lowering the likelihood of future wildfires.

The Western Fire Chiefs Association's inquiry, conducted in cooperation with Attorney General Lopez and other relevant parties, deepens and refines our comprehension of the Maui wildfires and the systemic problems that exacerbated their intensity.

The Historical Background of Maui's Fire Danger Warnings and Emergency Preparedness

The August 8, 2023, wildfires on Maui were not an isolated incident; rather, they were the result of decades' worth of environmental elements, land management techniques, and historical precedents. Understanding what happened on that fateful day requires an understanding of the historical background of Maui's fire danger alerts and emergency planning. Like many other places vulnerable to wildfires, Maui has long struggled to balance the needs of safeguarding its natural beauty with the needs of its people against the destructive power of fire. Throughout the years, initiatives to reduce the likelihood of wildfires have included vegetation management, prescribed burns, and public awareness campaigns about fire safety.

The possibility of wildfires has persisted as a concern for both officials and citizens of Maui despite these efforts. The island's distinct topography, with its untamed landscape and varied ecosystems, creates difficult obstacles for efforts to control and suppress fires. In light of this, fire danger alerts have been extremely important in advising locals about the possible threats posed by shifting weather patterns and environmental variables.

These warnings, which range from neighborhood notifications from Maui County's emergency management agencies to red flag warnings from the National Weather Service, are an essential tool for guiding preparation efforts and influencing decision-making.

But as the tragic events of August 8, 2023, showed, the success of fire hazard warnings depends on a planned and proactive reaction from all parties involved. Following the Maui wildfires, there is a newfound urgency to reevaluate current procedures, improve agency collaboration, and build community resilience against future threats.

By examining the historical context of fire danger warnings and emergency preparation on Maui, we gain valuable insights into the challenges and opportunities for building a more resilient and fire-safe island community. As we confront the realities of climate change and the increasing frequency and intensity of wildfires, we must draw upon the lessons of the past to shape a more secure and sustainable future for Maui and its residents.

Chapter two

Timeline of Events

Conditions and Warnings Before a Fire Several meteorological and environmental causes preceded the August 8, 2023, Maui wildfires, creating favorable circumstances for calamity. Maui had exceptionally high temperatures and little humidity in the days preceding the catastrophe, which increased the risk of flames around the island. These circumstances, together with blustery winds and parched vegetation, created the ideal conditions for the fire that would eventually consume the area.

Authorities warned people of the increased risk of fire hazard as a series of warnings were issued as the threat of wildfire loomed. Red flag warnings are issued by the National Weather Service to indicate that there is a greater chance of intense fire weather. Large portions of Maui were included in these warnings, and the Lahaina region was singled out for particular attention. This was an important reminder to both locals and government authorities to stay watchful and ready for the danger of wildfire.

The disaster management organizations in Maui County increased their readiness in response to these alerts, sending personnel to the county's emergency operations center and keeping a careful eye on any fire activity occurring throughout the island. Residents braced themselves for the likelihood of evacuation as the situation developed, but the full scope of the coming calamity remained uncertain despite these preventive precautions.

Initial Wildfire Activity and Evacuation Efforts

Early on August 8, 2023, reports of brush fires started to come in from various sections of Maui, marking the beginning of wildfire activity. Emergency personnel were notified of the existence of a fast-moving brush fire in the Kula area, some 20 miles from Lahaina, around 3:12 a.m. local time. As firefighters attempted to contain the fire, the Maui Emergency Management Agency (MEMA) responded by alerting Kula residents to leave their homes and find a safe place to retreat to.

Another fire was subtly burning in Lahaina, waiting to burst into a full-fledged firestorm that would soon engulf the town in a merciless swarm of devastation. What would subsequently be known as the Lahaina AM fire started at approximately 6:35 a.m. close to Ku-ialua Street and Ho'okahuaPlace. It quickly spread through the dry vegetation, endangering adjacent residences and businesses.

Emergency responders rushed to the scene as the fire grew more intense, sending firefighting crews and other resources in an attempt to put out the flames. As the fire spread into populated areas, locals were advised to leave; others were forced to leave their houses with little more than the clothes on their backs.

Events of the Day, Including Flare-ups and Fire Containment

On August 8, 2023, as flames raged and changed throughout the day, overwhelming emergency personnel and posing a serious threat to containment efforts, the situation on Maui became more and more dire. Firefighters used bulldozers and water tankers in a last-ditch effort to gain control of the Lahaina AM fire, fighting nonstop to create perimeter lines and saturate the fire area with water.

The constant onslaught of strong winds, however, quickly put an end to their efforts as they fanned the flames and sent flaming embers spiraling through the air, starting new fires in their wake. Reports of fallen power lines and utility poles in Lahaina started to surface by mid-morning, which made evacuation attempts much more difficult and made combating the fire more difficult.

The situation worsened throughout the day, with several flare-ups recorded around the region. Fueled by persistently high winds and dry conditions, the fires continued to spread unchecked despite the best efforts of emergency responders. Hundreds of residences and businesses had been consumed by the Lahaina fire by early afternoon, leaving a path of devastation in its wake.

Obstacles Emergency Responders Face Because of Strong Winds and Inadequate Communication

Emergency responders had numerous difficulties during the chaos and destruction caused by the Maui wildfires, which made it more difficult for them to manage the fire and safeguard towns that were at risk. The main obstacle in this situation was the strong winds that blew across the area, spreading the flames and embers across great distances.

With gusts as high as sixty miles per hour, these windy conditions put firefighters in danger and made it difficult for them to set up efficient containment lines. Power lines have occasionally entangled fire apparatus, making response operations more difficult and endangering first responders.

Communication breakdowns that hampered emergency response efforts all day long compounded these difficulties. Large portions of the island were consumed by the fires, overloading communication networks and making it impossible for emergency personnel to coordinate their efforts and get vital information to those in danger.

Emergency responders persevered in the face of these challenges, putting their lives at danger to save others and shield their towns from the flames' devastation. The tenacity and bravery of people who face nature's wrath in the course of their duties are demonstrated by their ceaseless efforts and steadfast dedication in the face of difficulty.

Chapter three

Response and Coordination Efforts

Maui Emergency Management Agency (MEMA) and County Officials' Reaction

Following the August 8, 2023, Maui wildfires, county officials and the Maui Emergency Management Agency (MEMA) moved quickly to contain the situation by organizing response operations and mobilizing resources. Entrusted with protecting the lives and assets of Maui residents, MEMA and county officials encountered an unparalleled difficulty in managing the swiftly expanding wildfires and intensifying situation.

As soon as the wildfires were discovered, MEMA triggered its emergency operations center, which brought together important players from different agencies to plan the response and share vital information. Together with MEMA, county officials—including the mayor and senior emergency management staff—assessed the situation on the ground and carried out evacuation preparations to protect the safety of locals in the impacted areas.

MEMA and county officials put in a lot of effort to get firefighting teams and resources to the front lines as the fires grew larger and more intense, mobilizing bulldozers, water tankers, and air support to put out the flames. Communities that were deemed vulnerable received orders to evacuate, and emergency shelters were set up to house the displaced

MEMA and county representatives kept in close contact with first responders during the crisis, offering them direction and assistance as they fought the fire and attempted to safeguard areas that were at risk. Their concerted efforts, in spite of the disaster's overwhelming scope, contributed to the preservation of life and limited further loss of property.

Requests for Extra Help and Interaction with State Organizations

MEMA and county officials realized they needed more help to support their response operations and contain the quickly growing flames as the size of the Maui wildfires became more apparent. Requests for extra firefighting resources, aerial support, and logistical aid were made to state agencies and other jurisdictions in order to bolster firefighting efforts.

Keeping in touch with state organizations, such as the Hawaii National Guard and the Hawaii

Emergency Management Agency (HI-EMA), was essential for organizing response activities and obtaining more resources to put out the wildfires. In order to guarantee a well-coordinated and efficient response, MEMA and county officials shared vital information regarding fire activity, evacuation orders, and resource requirements on a frequent basis.

But even with their best efforts, calls for further help were greeted with obstacles because of the state's limited resources as a result of other wildfires that were burning at the same time. Response operations were hampered by logistical issues and delays in the distribution of resources, underscoring the necessity of better resource management and coordination in future crises.

Resources and Coordination Limited in the Face of Increasing Fire Situations

The Maui wildfires of August 8, 2023, provided tremendous obstacles that taxed resources and tested coordination efforts amid worsening fire conditions, despite the coordinated efforts of MEMA, county officials, and first responders. MEMA and county officials faced the difficult issue of prioritizing response efforts and deploying resources where they were most needed as the fires spread quickly across the island, taxing firefighting capabilities to the breaking point.

Containment attempts were hindered by a lack of resources, including firemen, aircraft, and equipment, which forced them to make difficult choices about where to use what little they had. High winds and poor visibility hindered aerial support, which reduced the efficacy of aerial firefighting operations.

Coordination efforts were also severely hampered by communication breakdowns, as damaged infrastructure and congested communication networks made it difficult for agencies and first responders to exchange vital information. The already difficult scenario was made more difficult by delays in information relaying and response effort coordination, underscoring the necessity of robust communication systems and redundancy measures for such situations in the future.

Despite these obstacles, MEMA, county authorities, and first responders worked nonstop to lessen the effects of the Maui wildfires and save lives in the midst of the mayhem and devastation. Their commitment and tenacity in the face of difficulty are evidence of the Maui community's fortitude and resolution throughout difficult times. Building a more resilient and prepared community for the future will depend heavily on efforts to improve coordination, resource management, and communication as lessons are gained from the response to the wildfires.

Chapter four

Impact and Aftermath

The Lahaina Fire's devastation, which included fatalities and property damage

The Lahaina fire, which was the deadliest wildfire in the United States in more than a century, destroyed a great deal of property and caused a great deal of destruction in its wake. On August 8, 2023, when the unrelenting fire tore through the town of Lahaina, its citizens were forced to deal with a terrifying situation as their homes, shops, and landmarks were destroyed.

With 101 people losing their lives in the fire, it was one of the worst periods in Maui's history. The sad death of loved ones tore apart families, broke communities, and permanently changed the fabric of the island's close-knit community. The extent of the property devastation caused by the fire was equal to the human tragedy's magnitude.

Only burned debris and memories of past lives remained after thousands of houses and businesses were reduced to ash. Historic Waiola Church and LahainaHongwanji Mission, two landmarks that had existed for generations, were destroyed by the fire, their rich legacy lost to the destructive power of nature. The calamity had an equally catastrophic economic consequence, costing the local economy billions of dollars in losses.

When residents of Lahaina returned to their neighborhoods to assess the damage and start the arduous process of reconstructing their lives, the true magnitude of the destruction became brutally evident. The fire's scars would act as a menacing reminder of both the incredible power of nature at play and the frailty of human existence.

Firefighters' Struggles Fighting the Quickly Spreading Flames

Fast-moving flames fuelled by strong winds and dry weather presented a hard challenge to the firefighters fighting the Lahaina fire. Firefighters faced a number of difficult obstacles as the fire grew, putting their training, fortitude, and resolve to the test.

The overwhelming size of the fire, which swiftly outgrew containment measures and firefighting team capabilities, was one of the most urgent challenges. Driven by gusting gusts that transported blazing embers over great distances and started new fires in their wake, the flames spread alarmingly quickly. The difficulties firefighters encountered on the front lines were made even more severe by logistical problems and interruptions in communication. The exchange of vital information between agencies and first responders was impeded by overloaded communication networks and damaged infrastructure, making it challenging to plan response actions and adjust to quickly changing circumstances.

Firefighters persevered in spite of these challenges, putting their lives at danger to save others and shield their neighborhoods from the flames' devastation. Amid the turmoil and devastation, their courage and fortitude in the face of hardship gave hope and inspiration.

Attorney General Anne Lopez and Mayor Richard Bissen of Maui County made statements.

Following the Lahaina fire, Attorney General Anne Lopez and Maui County Mayor Richard Bissen released comments detailing their respective roles in the response and recovery operations, as well as offering their thoughts on the tragic event. Mayor Bissen offered his condolences to the families and communities impacted by the tragedy and expressed his deep sadness at the loss of life and property brought on by the fire. He promised to work nonstop to rebuild and recover from the destruction caused by the inferno, emphasizing the value of solidarity and fortitude in the face of misfortune.

Echoing Mayor Bissen's remarks, Attorney General Lopez emphasized the need for accountability, transparency, and justice in the wake of the catastrophe. She promised to look into the fire's origins thoroughly and to hold anyone accountable for any carelessness or misconduct that might have added to the tragedy.

Both leaders reaffirmed their pledge to assist the impacted communities and see to it that the catastrophe's lessons are used in order to avert such catastrophes in the future. Their words inspired hope and fortitude in the midst of catastrophe, acting as a rallying cry for solidarity and unity in the face of misfortune.

Federal agencies' ongoing legal actions and investigations

Following the Lahaina fire, numerous legal

proceedings and investigations were started in an effort to identify the causes of the catastrophe and hold people accountable for any negligence or misconduct.

With more than 135 individual plaintiff and class-action lawsuits filed in three separate courts, Maui County was left confronting an overwhelming volume of legal cases. These claims claimed damages for residents and companies injured by the fire, including money losses, property damage, and loss of life. Federal organizations, including as the US Bureau of Alcohol, Tobacco, Firearms, and Explosives (ATF), began looking into the origins of the fire in addition to the civil lawsuit. The goals of these investigations were to find any proof of arson or other illegal conduct that might have started the fire, as well as to pinpoint any institutional shortcomings in emergency response and preparation that might have made the situation worse.

Federal agencies promised to make public their conclusions as the investigations proceeded and to take appropriate measures to rectify any inadequacies or faults found in the fire response. Their actions served as a reminder of the necessity of learning from past mistakes to prevent similar catastrophes in the future, as well as the value of accountability and transparency in the wake of a calamity of this scale.

As locals struggled with the loss of life, property, and livelihoods, the full scope of the destruction caused by the Lahaina fire became brutally evident. The difficulties that firefighters encountered highlighted the critical need for enhanced reaction and readiness in the face of growing climate-related risks. Prolonged legal proceedings and investigations, along with declarations from local authorities, demonstrated the impacted communities' will to pursue justice and responsibility following the calamity.

Chapter five

Conclusion

An overview of the circumstances leading up to the wildfires in Maui and their effects

The August 8, 2023, wildfires on Maui serve as a sobering reminder of both the extreme destructiveness of nature and the significant influence of human choices during times of calamity. A path of devastation was left in its wake as what started out as a few seemingly harmless brush fires swiftly grew into an unprecedentedly large and destructive conflagration. With 101 fatalities and $6 billion in damages, the Lahaina fire in particular emerged as the deadliest US wildfire in almost a century.

A number of meteorological circumstances, such as high temperatures, low humidity, and strong winds, contributed to the conditions that were favorable for the start and spread of wildfires in the days preceding the Maui fires. Before the Lahaina fire broke out, officials disregarded extra assistance despite warnings from the National Weather Service about increased fire danger. This caused delays in response operations and increased the severity of the disaster.

Firefighters on Maui encountered numerous difficulties as the wildfires spread, such as strong winds, poor communication, and a lack of resources, which made it difficult for them to control the flames and save villages that were at risk. The unrelenting fire burned landmarks and reduced hundreds of homes and businesses to ash, resulting in an incredible loss of life and property.

Consequences for Upcoming Emergency Planning and Organizing

Policymakers, first responders, and communities around the country should review how they handle emergency planning and coordination in light of the growing risks posed by climate change in light of the Maui wildfires. The devastating effects of the Lahaina fire highlight the necessity of taking preventative action to lessen the likelihood of wildfires and guarantee a prompt and efficient reaction to calamities in the future.

Important takeaways from the Maui wildfires include the necessity of paying attention to early fire danger alerts, keeping up strong infrastructure and communication networks, and organizing response operations at all governmental levels. Specifically, turning down more assistance prior to the Lahaina fire emphasizes how crucial it is to be humble and receptive to outside assistance during emergency situations.

In the future, funding for vegetation management, community education initiatives, and prescribed burns will be crucial for lowering the danger of wildfires and safeguarding populations that are already at risk. Similar to this, strengthening response capabilities and guaranteeing a coordinated and successful response to emergencies will depend heavily on advancements in communication technologies and interagency cooperation.

The significance of comprehending the disaster's causes and lessons learned

In order to avert future catastrophes of a similar nature, we need to assess the causes and effects of the devastating Maui wildfires as the dust begins to settle. Comprehending the fundamental reasons of wildfires, including human choices and environmental elements, is imperative for formulating focused approaches to reduce the likelihood of subsequent calamities.

Learning from the errors and shortfalls of the Maui wildfire response is equally crucial in order to enhance future disaster planning and coordination efforts. Through comprehensive inquiries and disseminating insights to relevant parties, we can construct a more robust and flexible framework for handling crises and safeguarding localities.

The Maui wildfires ultimately serve as a sobering reminder of the pressing need for coordinated action to address the growing hazards posed by natural catastrophes and climate change. We cannot hope to create a future that is safer and more resilient for everyone unless we work together, be innovative, and make a commitment to proactive risk management.

www.ingramcontent.com/pod-product-compliance
Lightning Source LLC
Chambersburg PA
CBHW070950220526
45471CB00007B/2964